EXPLORE OUR NATIONAL PARKS

MR. BIGHORN'S
YOSEMITE ADVENTURE

BY DR. MIKE KOZUCH

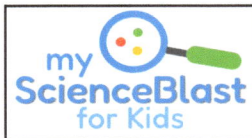

Title: Mr. Bighorn's Yosemite Adventure
Author and Photographer: Michael Kozuch
Publisher: myScienceBlast
Imprint: myScienceBlast for Kids
Website: myscienceblast.com

ISBN: 979-8-9936955-0-1
Printed in the United States of America

Photo Credits:
Photos © Michael Kozuch except where noted.
1883 Lyell glacier photo: USGS/Israel Russell (public domain).
2015 Lyell glacier photo: NPS/Keenan Takahashi (public domain).
Bighorn Sheep image from Pixabay (royalty-free license).

First Edition published 2025
Revised edition published 2026

Dedicated to all the young explorers who
keep asking "why?"

SHALL WE BEGIN ?

You can see these hills from the Tioga Pass Entrance.

Hi there! My name is **Mr. Bighorn**. I live in the high mountains of Yosemite National Park. Come with me and I'll show you how this amazing place was shaped by the forces of nature and carved by ice!

Before we begin, let me tell you about myself. I am one of a few hundred bighorn sheep that live in the **Sierra Nevada** mountains. We live on rocky cliffs to avoid predators. That means that we are great mountain climbers! Ok, let's get started on our tour.

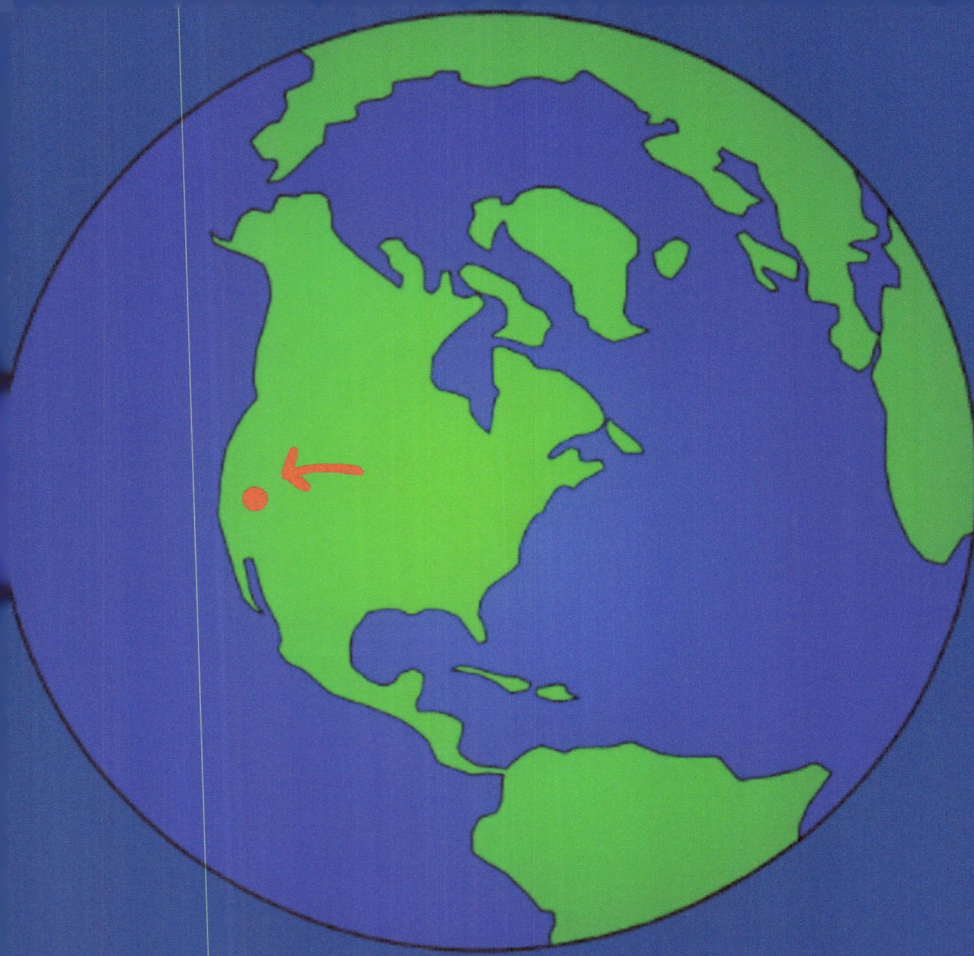

Some of you may be asking, "Where is Yosemite?"

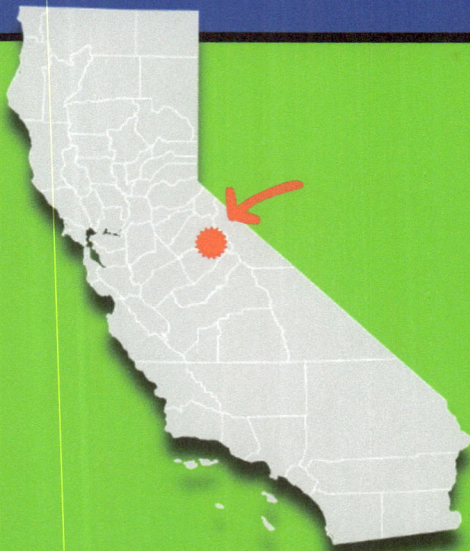

Yosemite National Park lies on the western side of the Sierra Nevada mountains in east-central California. It was officially named a national park in 1890.

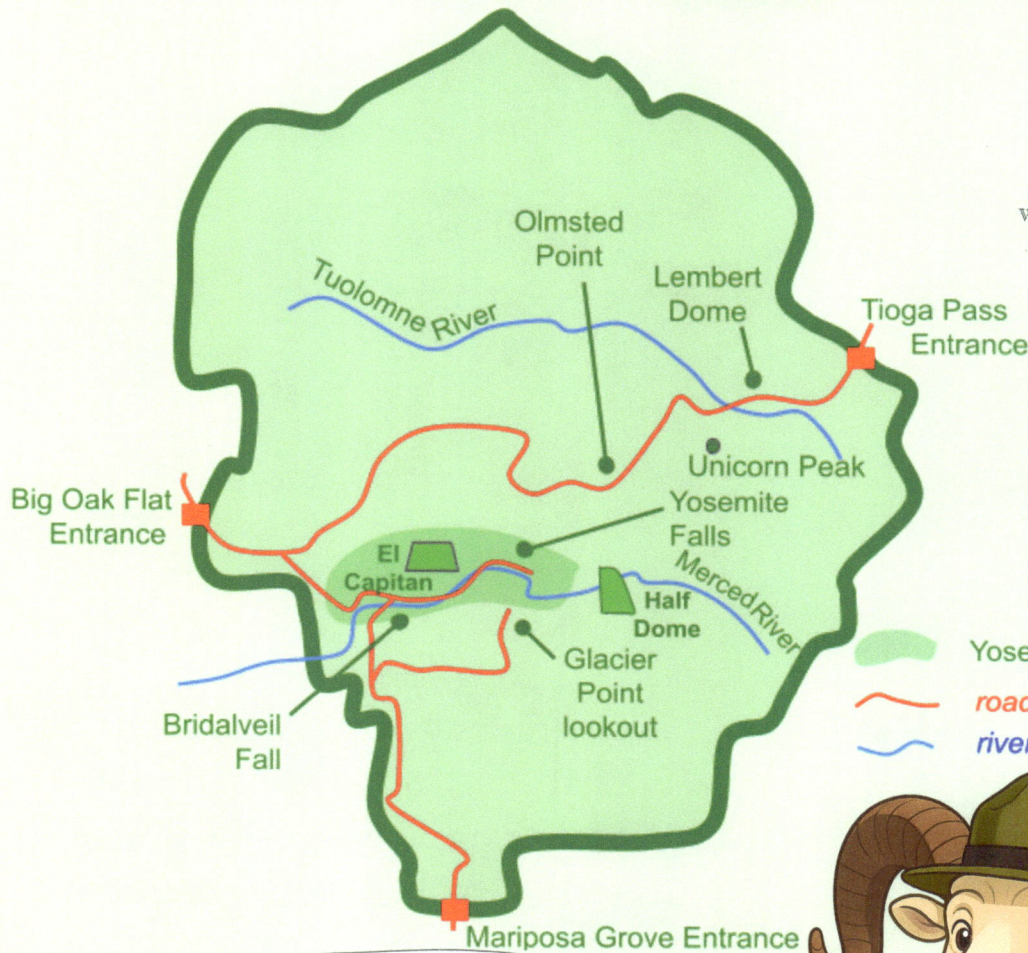

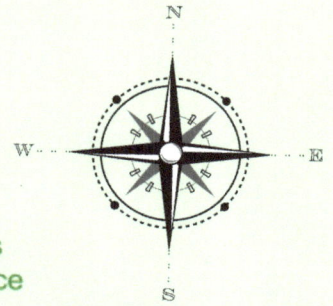

Yosemite National Park

Olmsted Point

Lembert Dome

Tuolomne River

Tioga Pass Entrance

Unicorn Peak

Yosemite Falls

Big Oak Flat Entrance

El Capitan

Half Dome

Merced River

Glacier Point lookout

Bridalveil Fall

Mariposa Grove Entrance

Yosemite Valley

roads

rivers

It's always a good idea for young explorers to know where they are going. You can use this map to keep track of all the places we talk about when visiting the park.

YOSEMITE

Where did all these rocks come from?

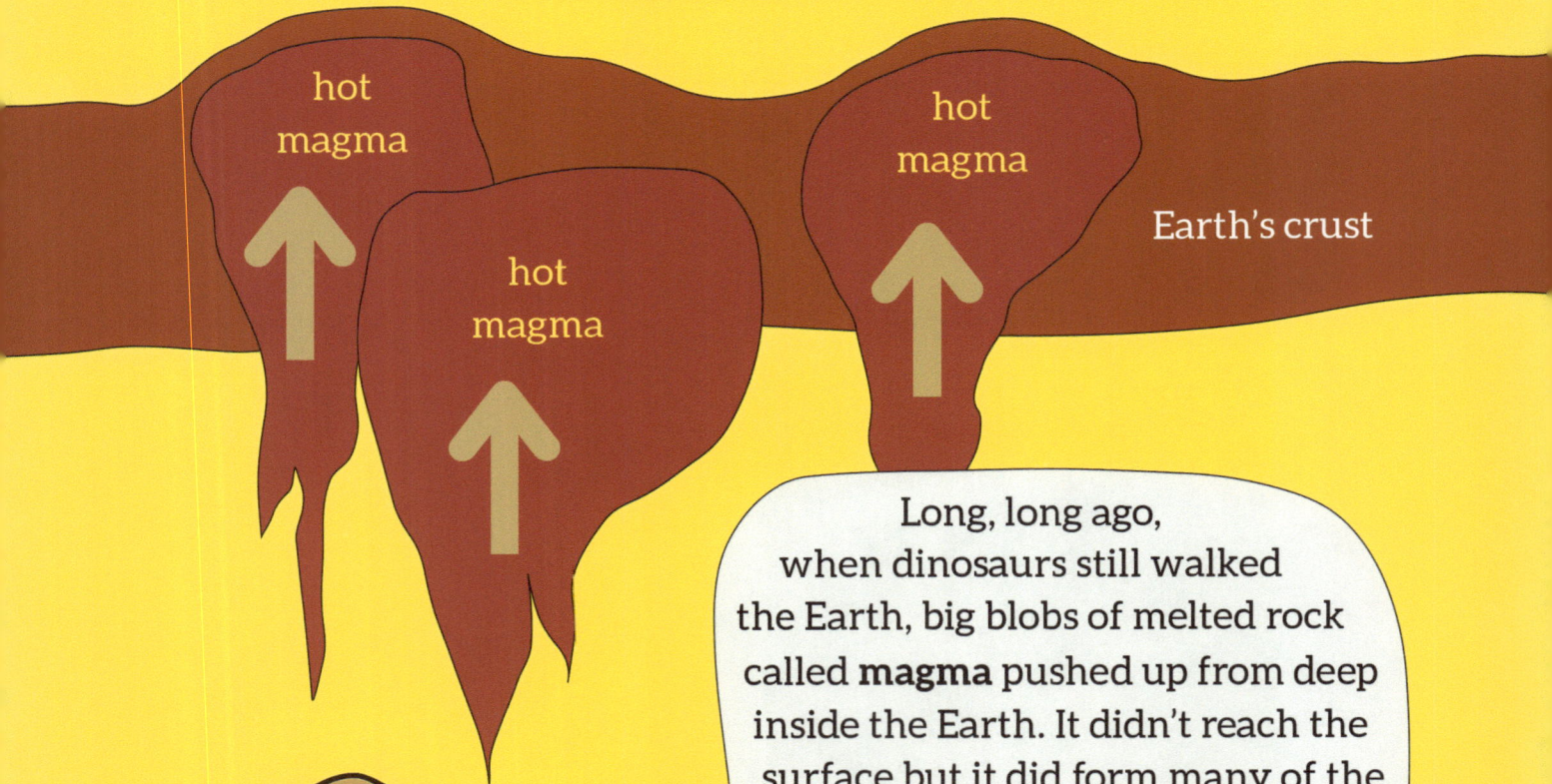

hot magma

hot magma

hot magma

Earth's crust

Long, long ago, when dinosaurs still walked the Earth, big blobs of melted rock called **magma** pushed up from deep inside the Earth. It didn't reach the surface but it did form many of the mountains you see. Eventually, the magma cooled and slowly hardened underground, forming giant blocks called **plutons**.

Millions of years later, wind and rivers wore away the dirt and rock that covered those plutons... and that's what you see today.

When you first come into **Yosemite Valley** you will see a lot of giant gray cliffs. That's **granite**, a super-strong type of rock which is what makes up the plutons. Granite rock is also made of shiny bits of something smaller called **minerals**.

Scan here to see how the hot magma forms these rocks!

What are these rocks made of?

granite rock

hornblende is a long dark mineral

quartz is a gray and glassy mineral

feldspar is a white mineral

If you pick up a piece of granite, you'll notice that there are smaller parts with different colors called **minerals.** Minerals are the ingredients of a rock, and **crystals** are the shape those ingredients make.

Taller than the Empire State Building

This tall granite wall is called **El Capitan**. It's over 3,000 feet high and if you look closely you can see a shape that looks like a heart. Can you find El Capitan on the map?

Watch the video on how El Capitan was formed and see people climbing this wall!

Next came the **glaciers**, thick layers of ice that covered the area a million years ago. Yosemite's glaciers acted like giant bulldozers. They scraped away loose rocks, polished the granite smooth, and carried huge boulders for miles. Sometimes they left big rocks behind, called **erratics**, that look as if they were dropped by giants. Erratics can be the size of a car! Have you ever seen one?

What did the glaciers do?

Look for erratics at Olmsted Point!

Check out the video on the awesome power of ice!

Here you can see a great example of how glaciers polished the granite when they passed over it. If you look for this pothole at **Lembert Dome** you will see a lot of smooth **glacial polish** nearby.

glacial polish

See that sharp ridge? That's **Unicorn Peak**! Glaciers carved both sides of it, leaving a thin, pointy edge called an **arête** (say: ah-RETT). It's like a mountain made by ice!

Glaciers also made the valley into a U-shape with steep sides and a wide, flat floor. That's why Yosemite Valley looks different from the sharp "V" shape that rivers make. When you stand at **Glacier Point** and look down, you can see the "U" clear as day.

1883

Lyell Glacier

2015

Glaciers don't last forever. A few hundred years ago, Yosemite had over 1,000 glaciers. Now only a few tiny ones remain like Lyell and Maclure Glaciers. They are shrinking because our climate is warming.

Bridalveil Fall

Yosemite Falls

Waterfalls are another gift from glaciers. Sometimes a smaller glacier met a bigger one. Since the larger glacier was more powerful, it eroded its valley much more deeply than the smaller one. This is why you sometimes see a small valley that hangs high above the larger valley floor. Today, rivers flow out of these "hanging valleys" as waterfalls. **Bridalveil Fall** and **Yosemite Falls** are some of the tallest in North America!

Test Your Knowledge

We've learned a lot! Let's see how much you can remember.

1. What do we call the big chunks of cooled magma that form mountains in Yosemite?
2. What 3 minerals is granite made of?
3. What is a glacier and what valley shape does it form?
4. What is an erratic rock?
5. What do you think Yosemite will look like millions of years from now?

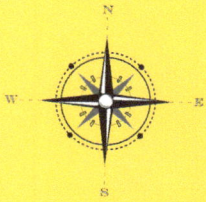

Yosemite Geology Word Search

Here is your chance to see if you can spot all the new words you learned in a Word Search. Circle the groups of letters that spell the words you find below the table. The words can go across, down, or even diagonal (a bit trickier).

Photocopy this page and give it a go!

D	G	L	A	C	I	E	R	A	P	Z	E	H	C
G	R	R	N	F	R	R	A	Q	L	F	T	S	L
R	E	M	K	T	D	E	A	O	U	R	C	E	I
A	O	N	N	I	L	O	L	N	T	N	C	O	F
N	P	C	L	C	F	I	M	D	O	I	I	M	F
I	R	O	K	R	P	L	A	E	N	N	A	S	F
T	E	R	O	S	I	O	N	O	A	E	E	E	E
E	V	R	E	N	E	R	R	A	T	I	C	T	L
Q	E	L	N	S	D	L	I	D	F	H	Y	P	D
U	L	O	T	S	P	R	L	C	L	O	C	I	S
A	T	F	S	P	O	L	I	S	H	R	T	I	P
R	A	H	P	R	C	R	Y	S	T	A	L	B	A
T	V	A	L	L	E	Y	O	N	G	N	E	E	R
Z	M	S	N	N	G	A	I	D	E	R	I	A	H
R	H	O	R	N	B	L	E	N	D	E	N	I	K

Words to Find:

GLACIER	ERRATIC	PLUTON	GRANITE	CLIFF
QUARTZ	FELDSPAR	HORNBLENDE	DOME	POLISH
VALLEY	EROSION	ICE	ROCK	CRYSTAL

ABOUT YOUR AUTHOR

Dr. Mike Kozuch is a geologist with many years of experience studying the geology of different parts of the world. He has taught a variety of courses in geology and oceanography at the university level and is the author of several university textbooks as part of his *myScienceBlast* series. He now writes children's books on these subjects to stimulate interest in these fascinating topics. Here you see him giving a lecture in Canyonlands, Utah.

www.ingramcontent.com/pod-product-compliance
Lightning Source LLC
Chambersburg PA
CBHW040811300326
41914CB00065B/1493